I Love Hugs and kisses

We all love hugs, and lots of animals love hugs too!

When you give a hug, you get one right back.

That's why hugging makes cuddly creatures feel safe, relaxed, warm, and happy.

We love to hold
each other **tight.**
We play fight and
then we **hug**
to show we are
still **friends.**

Baby elephants love
to play. They pretend
to fight and enjoy the
rough-and-tumble of playtime.

This is my dad's big belly hug! I love his warm and cozy cuddles.

Emperor penguins live near the South Pole and have to survive snow, ice, and freezing winds. A chick sits on its father's feet. His fluffy belly acts like a blanket to keep the chick warm.

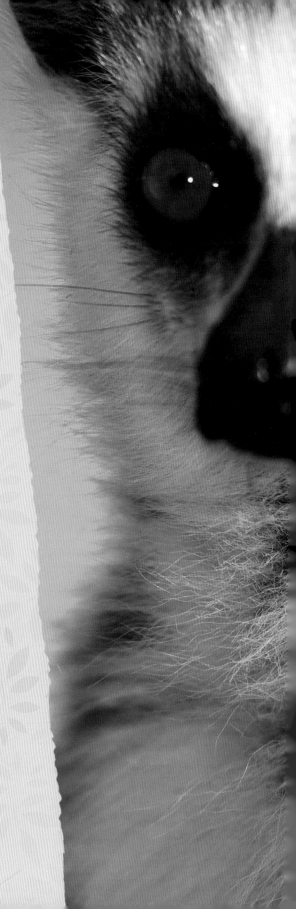

I love to **hug.**
I feel safe
when I hug
my **mom.**
I know she will
look after **me.**

When a lemur baby is born it holds
on to its mother's belly. As the
baby gets stronger, it can hold
on to its mother's back when
she leaps between trees.

The best hugs are big bear hugs.

A polar bear mother normally has two cubs at a time, and she will fight to protect them. The cubs play together to exercise, and to learn how to hunt.

We love
to cuddle
at bedtime.
It's easy to
fall asleep
when you
know someone
loves you.

Sea otters spend most of their
lives at sea, and they often hold
hands when they sleep. Baby otters
are called pups and their mothers
hold them on their tummies.

When a friend is feeling sad, just wrap your arms around them and gently squeeze.

Gibbons have very long arms. They use them to swing through the trees, and can travel 30 feet in just one swing. Their long arms are also perfect for giving comforting hugs.

The world isn't so scary when you've got someone to hug.

Meerkats live in groups of 20 or more called clans. They live underground in burrows but come up to the surface to look for food. Meerkats stand guard near their burrows and look out for danger.

We love to hug trees! We are brothers, and holding hands makes us feel all warm inside.

Mother raccoons have babies once a year. They have three to seven babies at a time, and they look after them in dens they have built in trees.

My dad gives me cuddles. It's his way of showing me just how much he cares.

A father lion looks after his cubs when the mother goes hunting. The cubs often pester their dad, but he gently pushes them away if they get too annoying!

We love to **snuggle** up.

We call this hug a **buddy body blanket!**

It is often cold and snowy in the places where Japanese macaques live. They keep warm by bathing in natural hot water springs, and by cuddling!

I know my mom loves me because she wraps her arms around me and tells me so.

Chimps are our closest relatives, so it is no wonder they like to hug as much as we do. They also like to play, and to be tickled and kissed!

Mom's asleep, but I can still sneak in for a quick snuggle...

Taking care of newborn tiger cubs is hard work! Cubs are born blind, so their mother must do everything for them. Tiger cubs begin to hunt when they are six months old.

We love a group hug! Everyone can join in...

Orangutan babies stay with their mother until they are about ten years old. Baby orangutans love to play, hold hands, hug, and kiss.

A hug is worth a thousand words. We don't need to talk when we hold each other tight.

When newborn pandas are born they are as tiny as an apple. A cub feeds on its mother's milk to grow. Later it will learn how to find and eat bamboo.

We love kisses,
and lots of animals
love kisses, too!

When someone kisses
you, it makes you
feel warm inside.

Animals kiss their
friends and their family.

A kiss is a wonderful way
to say hello or good-bye
and to show someone
that you love them.

Love can start with a kiss. A little peck on the cheek shows someone that I like them.

When a male little owl wants to find a mate, he calls out with a soft "hoot." A pair of little owls often stays together for life, and they both take care of their chicks.

There is **always** time for a quick **Kiss** when I say **good-bye.** I'll be back after **playing** with my **friends.**

Bottlenose dolphins live in groups called pods. Newborns stay with their mother. When they are bigger, all of the young dolphins in the pod play and swim together.

Sometimes Mom feels sad or tired. We keep kissing her until she feels happy again!

Cheetahs are good mothers. They work hard to protect their cubs from hyenas, lions, and eagles, and to find food for the whole family.

When my family comes over, I get smothered with kisses.

Zebras live in family groups called herds. Aunts, cousins, and sisters often help take care of a baby zebra, called a foal. They like to make a fuss over a foal and nuzzle and nibble its face and neck!

I'd like to get to know you better, and a little kiss is my way of saying "hello!"

Tree frogs are busy at night, looking for food. If two strangers bump into each other, they touch noses. It's a good way to decide if they can become mates.

It's bathtime, and Mom is trying to clean me.

She gives me a kiss to make me sit still!

Gorillas love to spend hours grooming one another. A mother checks every part of her baby's body to make sure that there are no nasty bugs living on his or her skin.

MOm gives me a kiss before she leaves. When she comes back, she will have some tasty treats for me!

Harp seal pups have fluffy, white fur and cannot swim until they grow short, gray fur. Their mothers must leave them alone on the snow while they go off fishing.

It's time to go. Kiss me quick!

Impalas are antelopes from Africa. Even when they are sharing a special moment, they must stay alert and watch out for lions or other predators.

Sometimes we fight, but we always kiss and make up.

Rabbits, like many other animals, use smell to decide if another animal is a friend or an enemy. Touching noses and mouths are just two of the many ways that animals can show that they trust one another.

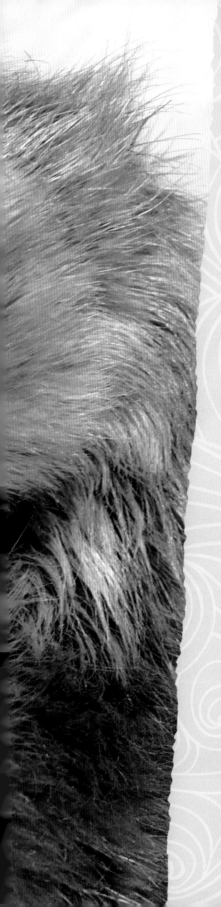

Dad has taken us **swimming**, and we are so tired! There's time for a sisterly kiss before we take a little nap.

Swan chicks are called cygnets, and their dad helps take care of them. He takes them to the river so that they can practice swimming. He might carry them on his back as he glides gracefully through the water.

I'm small and scared, but my mom's kisses make me feel big and brave.

A grizzly bear cub needs its mother. She feeds it and protects it from wolves, cougars, and other bears. A mother grizzly is a very dangerous animal when her cub is threatened.

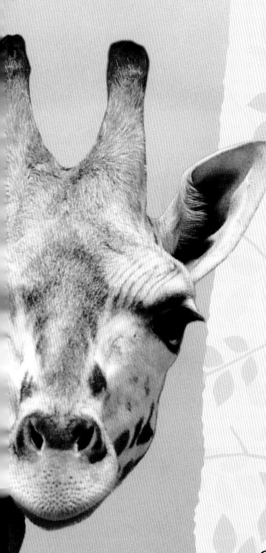

A kiss on the lips is lovely, but sneaky kisses are just as sweet!

A giraffe calf is about 7 feet tall and can run within one hour of being born. One mother takes care of a group of calves in a nursery, while the other mothers find food.

A kiss cheers me up when I'm feeling sad.

A pair of lovebirds spend their whole lives together. They like to sit close together and even nibble each other's beaks to show that they care!

Best friends love to share things, and a kiss is the perfect gift to give to that special someone.

Hippopotamuses are sociable animals. They live together in groups and often spend the day wallowing in pools to keep cool. At night, they leave the water to feed on grass.

Here are three good
reasons for Kissing.

A kiss is the shortest distance
between two friends.
♥
Kisses are easy to give and
lovely to get.
♥
A kiss is a gift of love.

Can you think of any more?

And three good
reasons for hugging.

A cuddle is a good way to show
someone you love them.

A snuggle makes you feel
safe and warm.

A hug is worth a thousand words...

Can you think of any more?

An Imprint of Sterling Publishing
387 Park Avenue South
New York, NY 10016

ISBN 978-1-4351-5942-6

Manufactured in Guangdong, China
Lot #:
2 4 6 8 10 9 7 5 3 1
11/14

Picture credits
(t=top, b=bottom, l=left, r=right, c=centre, fc=front cover)
1c: naturepl.com: Eric Baccega, 2c: Foto Natura: Flip De Nooyer, 3r: Shutterstock:
Background, 4l: Shutterstock: Background, 5r: naturepl.com: Tony Heald, 6c: Frans
Lanting Stock: Frans Lanting, 7r: istockphoto.com: Background, 8l: Shutterstock:
Background, 9r: naturepl.com: Anup Shah, 10c: FLPA: ImageBroker, 11r: istockphoto.
com: Background, 12l: Shutterstock: Background, 13c: FLPA: Suzi Eszterhas, 14c:
FLPA: Christian Hütter, 15r: Shutterstock: Background, 16l: istockphoto.com:
Background, 17r Getty Images; (c) Paul Souders, 18c: Minden Pictures: Tim Fitzharris,
19r: Shutterstock: Background, 20l: Shutterstock: Background, 21c: Minden Pictures:
Suzi Eszterhas, 22c: Foto Natura: Stephen Belcher, 23r: Shutterstock: Background, 24l:
istockphoto.com: Background, 25r: Corbis: © Frans Lanting, 26l: FLPA: Suzi Eszterhas,
27r: Shutterstock: Background, 28l: Shutterstock: Background, 29r: FLPA: Mitsuaki
Iwago, 30l: ZSSD, 31r: Shutterstock: Background, 63c: Shutterstock: Background, tl:
© Biosphoto: Michel & Christine, 32 FLPA: Minden Pictures: Gerry Ellis, 33
Shutterstock: Sweet Lana, 34 Shutterstock: Daemys, 35 Naturepl.com: Dietmar Nill,
36 Naturepl.com: Jeff Rotman, 37 Shutterstock: Markovka, 38 Shutterstock: EV-
DA, 39 FLPA: Frans Lanting, 40 FLPA: Richard Du Toit, 41 Shutterstock: yaskii, 42
Shutterstock: Sweet Lana, 43 Getty: JH Pete Carmichael, 44 Alamy: Dave Stevenson,
45 Shutterstock: WEN WEN, 46 istockphoto.com: Electric Crayon, 47 Getty: Hiroya
Minakuchi, 48 FLPA: Minden Pictures: Suzi Eszterhas, 49 Shutterstock: Natalia
Kudryavtseva, 50 Shutterstock: yaskii, 51 FLPA: Andrew Parkinson, 52 naturepl.
com: Paul Hobson, 53 Shutterstock: Markovka,
54 istockphoto.com: Silmen, 55 Biosphoto: Michael Breuer, 56 Getty: Gail Shotlande
57 Shutterstock: Gizele, 58 Shutterstock: Daemys, 59 Imagebroker: Michael Krabs,
60 Naturepl.com: Anup Shah, 61 Shutterstock: Natalia Kudryavtseva,
62 Shutterstock: Sweet Lana, 62t FLPA: Minden Pictures: Yva Momatiuk & John
Eastcott, fc Getty: Roland Weihrauch, bct Shutterstock: Berndt Vorwald,
bcbl Shutterstock: Moolkum, bcbr Shutterstock: David Steele